Barr Meadows

Eruptions

Their Real Nature and Rational Treatment

Barr Meadows

Eruptions
Their Real Nature and Rational Treatment

ISBN/EAN: 9783337022440

Printed in Europe, USA, Canada, Australia, Japan

Cover: Foto ©berggeist007 / pixelio.de

More available books at **www.hansebooks.com**

ERUPTIONS:

THEIR REAL NATURE, AND RATIONAL TREATMENT.

BY

DR. BARR MEADOWS, F.A.S.L.,

Lic. Royal Coll. Phys. Ed.; Mem. Royal Coll. Surg. Lond.; Physician to the National Institution for Diseases of the Skin; late of the Royal Pimlico Dispensary; and formerly in the service of H.M.'s East Indian Government.

FOURTH EDITION.

"Although the MEDICAL WORLD is under strong and lasting obligations to the late Dr. Willan, and to his pupil, Dr. Bateman, for their arrangement of Cutaneous Diseases, in which they have exhibited much ability and unwearied industry; very little has been added to our stock of knowledge regarding the cure of these most troublesome and vexatious diseases."—*Edinburg Medical and Surgical Journal.*—Vol. xvi.

LONDON:
GEORGE HILL, 154, WESTMINSTER BRIDGE ROAD.

1867.

PRICE, HALF-A-CROWN.

To Dr. R. GARDINER HILL,

Originator of the non-restraint system of treatment in lunacy.

My Dear Sir,—

The amelioration in the condition of the insane through the now universal adoption of the non-restraint system, was not, as you know, effected without much opposition; and, although the misrepresentation to which you were exposed in the outset of your crusade against the old, crude, and worse than useless treatment formerly adopted, has, doubtless, in the final triumph of your efforts and the improved state of psychological science been, in a great measure, forgotten, you will not, I feel certain, fail to sympathize with me in the struggle against, what I conscientiously believe to be, traditional errors.

Allow me, therefore, to inscribe your name upon my little work, and believe me,

Your affectionate Nephew,

THE AUTHOR.

PREFACE TO FOURTH EDITION.

In confirmation of my statements as to the empirical character of the system which is still taught and practised by the great body of the profession, I have thought it right to introduce, as foot notes, several passages from the more recent publications upon this subject—some of which, I am happy to think, give complete, if unintentional, corroboration and support to the views which it is my sincere wish to see generally and practically accepted. It is now some years since these views were published, and a largely increased experience continues to confirm my opinion as to the soundness and value of the simple, as compared with the specific systems. The general literature of medicine, likewise, affords ample evidence of the gradual, but still constant and progressive change which is taking place in the views of the profession, some, even amongst those to whom systems of classification are necessities, directly or indirectly, being forced to acknowledge their dependence upon general indications in the treatment of cutaneous

diseases, and, whilst maintaining their ideal speciality, to modify, somewhat, their faith in the practicability of specifics. · Whether the publication of this little book has in any way contributed to this alteration I do not presume to say—I notice its occurrence subsequently—but with every confidence in the eventful downfall of all artificial classifications of disease, and of all empirical systems, once more offer to the reader my reasons for repudiating the old and well-worn teachings of the schools.

BARR MEADOWS.

2, Warwick Street, Cockspur Street,
Charing Cross, Sept., 1867.

INTRODUCTION.

THE aim of the following pages is to demonstrate the symptomatic nature of eruptions generally, to endeavour to disabuse the mind of prejudices in favour of the unscientific and obstructive system of classification so universally in vogue, and to point out the one only, natural, simple, and successful plan of treatment, which, in accordance with the general principles of Medicine, may always be relied upon.

Without hesitation as to the expression of an opinion, as to the empirical character of all plans of treatment having regard only to objective symptoms, we, at the same time, desire to be distinctly understood, to have no intention to disparage, or fail in due respect to those gentlemen who follow the general practice or promulgate the general views ; they but follow the teaching they have received, and which they no doubt regard as authoritatively perfect. Many learned men have, indeed, given to the system originally promulgated by Drs. Willan and Bateman, all the support which any such method is capable to receive ; but no authorities, however celebrated, can make wrong right, or controvert reason and common-sense.

Willan's system was introduced at a period when physicians knew literally nothing about the nature or treatment of eruptions ; and, truth to say, but little more about diseases generally—empirical practice was then the rule, and no method, in consequence, was more calculated to be well received than the one in question—whilst supported by men of note and celebrity, it, in course of time, began to be looked upon as "an institution," the age of which, ever increasing, has maintained for it respect. Many other traditional fallacies have been swept away, rapid strides have been made in the various branches of Medical knowledge, and the collateral sciences, and we make a call to all concerned in a desire to improve the practice of Medicine, to dismiss from their minds all preconceived ideas as to the special nature of these disorders, and with unbiassed judgements, to decide between " differential diagnosis " and common sense—to allow reason to replace tradition, and to extend to affections accompanied by cutaneous eruption, those mere rational rules of practice, which a knowledge of their real character demands.

Cutaneous affections rank confessedly amongst the opprobria of our art. There are, indeed, sufficient *natural* difficulties in their treatment, but the chief, whereby they so frequently baffle the best efforts of the practitioner, is, we opine, of easy explanation, and to be found in that misapprehension of their nature, which originated in, and is fostered by, the false system of classification to which they have become subservient.

The earliest records of Medicine afford sufficient evidence that these disorders have always, and from time immemorial, been looked upon as peculiar and distinct ailments—diseases *per se*—requiring a special line of treatment; a view no doubt originating in their notorious resistance to the numerous remedies which, at various periods, have been unsuccessfully administered for their relief.

Prior to the advent of Willan's methodical arrangement, various attempts had been made by physicians, to classify skin "diseases," and, as they imagined, thus to render them subservient to scientific study. Mercurius, Plenck, and Alibert, were amongst the most celebrated for their systems of classification, all of which, it is worthy of remark, were founded upon the variations presented by the eruption, either in appearance, physical attributes, or situation. These earlier attempts are now almost forgotten, whilst the system initiated by Willan, and subsequently more fully developed and promulgated by his friend Dr. Bateman, has continued to be looked up to, and followed, not only by his own countrymen, but generally by practitioners throughout the world. Taught in our colleges, practised in our hospitals, and in universal favour amongst private practitioners, it has long taken firm hold upon the professional mind as a valuable addition to medical science, and this, too, in spite of the frequency with which in practice its failure is demonstrated, and made apparent.

It is no light matter to assail, and disallow a system in such universal favour, more especially when, as in the present instance, it is not intended to offer in its place an equally or more pretentious substitute. What!! after arriving at the dignity of a separate science; after gentlemen of standing and position in their profession have given up their attention solely to the study of "dermatology;" after the learned societies and journals of the profession have for generations been attentive to the praises of every new kind of remedy or special mode of treatment, which may have been vaunted in this or that form of eruption; when our hospital museums are ornamented with expensive wax models, and the windows and shelves of medical publishers are filled with plates, designed to illustrate the different forms which eruptions are wont to assume, and to assist the practitioner in his endeavour after correct " differential diagnosis;" after all this, to be told that cutaneous disorder is always, and only symptomatic, and should be treated upon the ordinary principles of Medical science, it cannot be!—yet such is the duty which we have undertaken to perform, nor do we imagine that it will be difficult, to convince any *unprejudiced* reader, that the simple, natural, and therefore rational plan we follow and would recommend, is far superior to any artificial nosological arrangement or theoretical system.

To those who have had experience of the difficulties attending the usual practice, and who must be only too well aware how frequently under its auspices the cutaneous disorder resists all the usual and routine efforts at cure, we commend a trial of the rational method, content to rest its claims upon the superior results that will inevitably ensue.

ERUPTIONS, &c.

Drs. Willan and Bateman, and all the more modern writers who have adopted their system, commence by the division of " skin diseases " into eight orders; the Papulæ, *pimples;* the Squamæ, *scales;* the Exanthemata, *rashes;* the Bullæ, *blebs or large vesicles;* the Pustulæ, *pustules;* the Vesiculæ, *vesicles;* the Tubercula, *tubercles;* and the Maculæ, *spots*. The pimples, scales, vesicles, etc., are the elementary lesions.

The next step consists in the division or subdivision of the eight orders into different genera; such division being still in accordance with, and in consequence of, the differences of appearance, and physical peculiarities, presented by the eruption. These again, and very frequently merely on account of some accidental difference in their location, are, in like manner, resubdivided into species.

To each genus, and to every variety in which such genus may be subdivided, there is attached some Greek or Latin, or *Greek and Latin* appellation, with which being dignified, it (the eruption) takes rank as a separate and independent disease. The minute and verbose descriptions of the appearance, physical peculiarities, and progress of the eruption that follow and belong to each specially named variety, it is not easy to describe, and those who are unacquainted therewith, and who may not care to lose their time in such researches, must consult the wordy, complex, and contradictory writings of the classical authors upon this subject; suffice it to say, that the hardness, dryness, locality, temperature, colour, sensations, &c., &c., &c., which are supposed to distinguish the several species are, in such works, descanted upon in a manner, and with a fulness, evincing an attentive observation and patient labour which would be most praiseworthy were it only useful.

Thus, mapped out, named, and described, dermatologists proceed to discuss the special line of treatment which they consider suitable to each variety, and to enumerate those remedies with which they have had the most success, or which have been recommended by other writers on the subject. Little do we find in such works touching the causes of the cutaneous disorder *in individual cases;* occasionally, we meet with acknowledgments of accompanying constitutional disorder, introduced in a general and incidental manner, or referred to, as symptomatic of the cutaneous " disease ; " usually, the causes are said to be "*various and obscure*," and so, together with the constitutional and general symptoms dismissed, to give the writer space to enlarge upon the several special remedies, which, in the natural relation of things, have come to be considered as necessary in the treatment of these cases.

The following extract from the work of Dr. Burgess will give the reader some idea as to the essentially special point of view from which cutaneous affections are regarded and of the stress that is laid upon " correct differential diagnosis."

" The differential diagnosis of diseases of the skin is " one of the most important points connected with their " history ; we shall, therefore, endeavour to lay down some " general rules for our guidance. The chief point is to " determine the elementary lesion ; this done, wo have " merely to compare the disease with the few which possess " the same elementary characters. In cases where the " elementary lesion remains unaltered, we have simply to " ascertain whether it be a papule, vesicle, scale, &c., and " this generally is a very easy task. Our next step is to " determine the species, and in this we are aided by the " *form, seat, progress,* &c., of the eruption.

" For example, a patient has, on the inner side of the
" arm, between the fingers, &c., a number of *small collec-*
" *tions of serum*, distinct, accuminated, transparent at the
" point, and accompanied by itching, &c. On carefully
" examining, we find that the elevations contain no pus,
" that they are not solid and resisting, that they are not
" papular eminences covered by a scale, nor an injection
" of the skin which disappears under pressure ; the disease
" is therefore *vesicular*. We have then to find out to what
" species of vesicular affection it belongs. It is neither
" *miliaria* nor *varicella*, which are accompanied by con-
" stitutional symptoms ; it is not *herpes*, for in herpes the
" eminences are collected together in groups; it must
" therefore be either eczema or scabies; but it is not
" eczema, for the vesicles of eczema are flattened, while
" here they are accuminated ; *ergo*, it is scabies.

" The example which we have just given is a simple one ;
" but the diagnosis is sometimes more difficult, even when
" the elementary character of the disease remains in part :
" thus scabies, which is generally detected with readiness,
" may sometimes present some difficulties of diagnosis,
" especially when the vesicles have been destroyed by
" scratching; but in such cases we are assisted by various
" secondary indications, such as the seat of the eruption,

"the appearance of its accidental variety, the precursory
"and accompanying symptoms.

"But a mere knowledge of the elementary character
"of a cutaneous disease is not sufficient for its diagnosis;
"this character may have disappeared, and given place
"to the secondary or consecutive lesions. The fluid of a
"vesicle may dry off and leave a small incrustation; a
"pustule may be converted into a scab, and the latter give
"way to an ulcer; hence it is necessary that we should
"study these secondary lesions, and know to what primary
"characters they correspond. Incrustations may succeed
"vesicles; scabs occur in most pustular diseases, and
"ulcerations may be a consequence of rupia, ecthyma, &c.

"In cases like the foregoing, we must first ascertain the
"nature of the secondary lesion, then determine its corres-
"ponding primary element, and finally pursue the course
"just pointed out. For example, a patient comes to us
"with a disease of the skin, characterized by thick, rough,
"yellow scabs, which cover a large portion of the
"extremities, especially the legs, and when they fall off,
"expose superficial excoriations; the latter discharge a
"purulent secretion, which dries up, and forms fresh scabs,
"these being the most characteristic features of the
"disease. Now it is easy enough to tell at once that this

" is a pustular affection, but not so easy to determine its
" species. The disease is evidently neither *variola* nor
" *vaccinia;* the pustules of *ecthyma* are large, isolated, and
" frequently covered by black, tenacious scabs, which end
" in ulceration; it is neither *acne* nor *mentagra*, the
" pustules of which rarely ever give rise to scabs. The
" only affections, then, that remain are *impetigo* and
" *porrigo*, and we have merely to compare the character
" of these two species in order to decide. It is unnecessary
" to enumerate here the signs by which we know that the
" disease is not porrigo; it is therefore impetigo, and as
" the scales are scattered irregularly over the limb, it is
" *impetigo sparsa.*

" In some cases different elementary lesions occur in
" the same subject; but even here we always find some
" predominent form, of which the rest are but complica-
" tions. However, it may happen that we cannot ascer-
" tain at once the true nature of the disease. This occurs
" in certain chronic affections, where the elementary
" character gradually disappears, and seems confounded
" in a different order of phenomena. Even here, a sudden
" exacerbation of the disease, or a return to health, may
" develop its primary character." *

---

* " Manual of Diseases of the Skin, from the French, by M. M. Cazenave and
Schedel, &c., by T. H. Burgess, M.D., Surgeon to the Blenheim Street Dispensary
for Diseases of the Skin," pp. 13-15.

For such painstaking "differential diagnosis" it is however, unfortunate, that in a given number of cases which shall present themselves for examination, no two will be found precisely similar, so that in any individual case objective appearances can only in an approximate manner resemble the descriptions to be found in books.

This will account for the fact that all dermatologists differ with regard to the characters which are supposed to distinguish particular genera or species, one gentleman being of opinion that a certain eruption should be called by this name, another gentleman that it should be recognized by that.

Hence, likewise, the ever fresh names which are constantly being introduced, and the various new species which are continually being recognized and classified. * †

This method, then, professes to allot the different forms of cutaneous disorder into orders, genera, and species,

---

* "No subject in the study of medicine has created more difficulty. or for a longer period tended to retard its advancement, than that of nosological arrangements." • • * "Writer after writer, impelled, as it were. by an ambition to devise something novel, propounds a classification. careless how complicated and difficult of being comprehended it may be, provided only it differs from those which preceded." Negligan on Skin Diseases, by Dr. Belcher, 2nd edition, p. 4.

† "There is probably no class of diseases less understood, both by medical students and practitioners, than the class of Skin Diseases. There are several causes which have conduced to this result. One cause is the great diversity of names which have been given to these diseases by different authors: some diseases having several names, and the same name having often been given to

and from the peculiarities supposed to characterize the
appearance of each, to endue such form or genus with a
distinct entity and appellation,   We shall not, in this
place, stop to enquire too closely as to the feasibility of
such a proposition, but allow, for the sake of argument,
that this the first step in the programme of authors upon
classification, is accomplished. What then? The patient
before us, and his skin exposed, presenting such well
marked, peculiar and abnormal appearances and physical
characteristics, that it were impossible but to place it to
its right order, and to recognize it by its right and
appropriate name.   Will such facility in " differential
diagnosis " enable us to appreciate the constitutional
peculiarities in our patient?   Shall we, by such means,
arrive at a knowledge of the *causes* engaged in the
production of the eruption?   Or shall we, simply, and
without such guides, from the designation and physical

---

diseases totally distinct from each other ; even the same writer has given new
names to diseases described previously by himself under other names, and
some authors use familiar terms with well recognized meanings in a manner
quite peculiar to themselves." "The difficulties of understanding skin diseases
have been increased not only by the multiplication of long names, but by endless
varieties of classification and extreme subdivision." (*Dr. Hillier's work contains
in the appendix a new classification of his own, which occupies three pages.*)
" From the fact that all morbid changes in the skin are open to inspection and
have been closely observed, the names of skin diseases have been multiplied, the
same disease receiving different names from the different appearances presented
by it at different stages of its progress, or from variations in its severity, or from
peculiarities in the individual." Hillier on Skin Diseases, 1865.

attributes of the cutaneous disorder itself, be in a position to judge as to the nature of the remedial measures requisite for the cure of any particular case? If the objective symptoms be incompetent to fulfil such office, then have we gained just nothing by our acquaintance with their *"distinctive"* characters as enumerated in the nosological tables of dermatology. Practically, this is conceded by those who follow Willan's method—for each form of eruption has, in fact, allotted to it, its own special remedy or remedies, which, according to the different authorities, either should or may prove beneficial. How far this theory is pushed, and how little it is borne out in the result, we shall leave the reader to ascertain; employing as evidence to elucidate this and other important facts connected with the subject, such extracts from the writings of both special and general medical authors, as shall seem most forcibly to bear upon the points severally to be discussed; premising, that such quotations are chosen fairly and impartially, and from the writings of gentlemen well qualified as authorities.

The following comprise some of the more serious objections which we purpose to consider in reference to the prevailing system.

I.—That the generic appellations by which cutaneous affections are designated have reference only to their physical and objective characters, and are, with one or two exceptions, totally irrespective of their constitutional relationship or causation.

II.—That the generic terms are used uncertainly;—a similar form of eruption being differently named by different writers; and *vice versa*, the same name being in use as indicative of dissimilar forms.

III.—The frequent impossibility of deciding as to which form mentioned in the nosological tables the case under consideration is referable.

IV.—The simultaneous presence in the same individual of several, more or less distinct, forms of eruption.

V —The alteration of the form of eruption from time to time in the same case.

VI.—The fact that one and the same cause may give rise to various forms of eruption.

VII.—That a similar eruption may depend upon many and very dissimilar causes.

VIII.—That the peculiar form presented by the eruption is practically useless as a guide to treatment.

IX.—The empirical and unsatisfactory nature of special plans of treatment, and of so-called specific remedies.

First, as to the generic appellations by which the different abnormal appearances presented by the skin are known and distinguished. This is necessarily a point of great importance when the method of treatment is supposed to be dependent upon the form of eruption, and, consequently, we shall not be surprised that Dr. Bateman evinced much anxiety for a vocabulary sufficiently copious to distinguish and describe his many forms and species.

"Amongst the manifest advantages of a copious and "definite nomenclature, may be mentioned in the first "place, the necessity which it demands, of an accurate "investigation of phenomena, or, in other words, the "habitual analytic turn which it tends to give to our "enquiries, and therefore the general improvement of the "talent of observation which it must ultimately produce. "Secondly, it contributes to facilitate the means of dis- "crimination, by multiplying, as it were, the instruments "of distinct conception: for from a deficiency of terms "we are apt to think and even to observe indistinctly. "But, above all, a definite nomenclature supplies us with

" the means of communicating with precision, the informa-
" tion which we acquire, and therefore contributes directly
" to the advancement of knowledge, or at least removes an
" otherwise insurmountable impediment to its progress.
" In this view, such a nomenclature, as far as regards the
" diseases of the skin, is obviously a great desideratum.
" For, while the language taught us by the fathers of
" medicine, relative to all other classes of disease, is clear
" and intelligible, the names of cutaneous disorders have
" been used in various acceptations, and without much
" discrimination, from the days of Hippocrates, and still
" more vaguely since the revival of learning in modern
" times. From that period, indeed, the diseases of the
" skin have been generally designated by some few terms
" of universal import, which therefore carried no import
" at all. Hence the words, Leprosy, Scurvy, Herpes,
" Scabies, Dartres, and some other appellations, have
" become so indefinite, as to be merely synonyms of
" cutaneous disease. Even the more scientific enquirers,
" whose knowledge of diseases was not always equal to
" their learning, or whose learning fell short of their
" pathological skill, have interpreted the generic and
" specific appellations of the ancients in various senses.
" They have not only differed, for instance, in their

" acceptation of general terms, such as the words pustule,
" phlyctæna, exanthema, erythema, phyma, phlyzacium,
" etc.; but the particular appellations, Lichen, Psora
" Herpes, Impetigo, Porrigo, Scabies, and many others,
" have been arbitrarily appropriated to very different
" genera of disease. The practical errors, which must
" necessarily have resulted from such a confusion in the
" use of terms, are very numerous, as every one must
" be satisfied, who has attempted to study the subject in
" books. It may be sufficient to allude to the gross
" misapplication of the remedies of the petechial or sea
" scurvy, which have been prescribed for the cure of
" inflammatory, scaly, and pustular diseases, merely
" because the epithet, *scorbutic*, have been vaguely assigned
" to them all; and to specify the single instance of the
" administration of tincture of Cantharides in the scaly
" Lepra, on the recommendation of Dr. Mead, who however
" seems to have spoken of the tubercular Elephantiasis, or
" the non-scaly Leuce; although it would be very difficult
" to ascertain his meaning. Most of the writers, who have
" composed express treatises on cutaneous diseases, in
" modern times, have implicitly adopted the nomenclature
" of the ancients, without attempting to render it more

"definite, or to improve upon the diagnosis which they
"had pointed out.* "

The above extract clearly demonstrates three of the
fundamental principles or peculiarities belonging to this
system; viz. :—the peculiarly distinct nature supposed to
attach to cutaneous disorders generally, and to the various
forms in particular; the special plan or plans of treatment
imagined to be requisite in each several form : [and the
entire absence of regard for variety in causation as a guide
to the choice of remedies in particular cases. Dr. Bateman's own method, being but an exaggerated example
of a like proceeding—his condemnation of those who
employ one remedy for very various forms of disorder,
"merely because of the epithet vaguely assigned to them
all," is specially amusing. Drs. Willan and Bateman did,
and those who follow their teaching do, exhibit one form of
remedy in many very different constitutional disorders,
merely because a single symptom (the eruption on the
skin) presents a like appearance in all, and has had a
perhaps by no means characteristic epithet assigned to it
by these very gentlemen.

Sailors, after a long and hazardous voyage, not infrequently arrive in port, in broken health, with bleeding

---

* Bateman on Cutaneous Diseases, 2nd edition, pp. 28-30.

gums, and bruise-like discolorations of the skin, etc.—in short, suffering from what is popularly known as sea-scurvy.

"The superficial markings of purpura, the red and purple spots and livid blotches *exactly resemble* the spots and bruise-like discolorations which characterise sea-scurvy,"* yet the causes and the constitutional condition in each are by no means the same, and, consequently, their successful treatment can only be conducted by different, and, frequently, contrary means.

The physical and abnormal peculiarities presented by the skin are, in such cases, sufficiently well marked, but they never did, and they never could *alone*, have afforded any information or guidance as to the remedial or preventive measures demanded for, and adapted to the malady with which they are connected.

From *the history of the causes* engaged in numerous cases, sprang the knowledge that want of fresh vegetables caused, and that their subsequent use would cure the disease ; and so, scientific enquiry having shown that potass was not only a natural constituent in healthy blood, but also, more or less, in most vegetable articles of

---

* Dr. Watson.

diet, we have an explanation of the cause and intimate nature of the disorder resulting from vegetable deprivation, and a sure and simple guide to its prevention or its cure.*

The small reliance that is to be placed upon the mere form assumed by the cutaneous symptoms as a practical guide to rational and successful treatment is well exemplified in this affection. A like form of cutaneous disorder to that which ensues upon vegetable deprivation, and even to a *certain extent* a similar constitutional derangement, known also under the generic name of *purpura*, is not infrequently met with amongst the poor, ill-fed, half-clothed, and houseless denizens of our larger cities;—these poor wretches have, may be, from dire necessity, existed for some lengthened period upon an insufficient and *vegetable* diet;—the outward and superficial markings are the same in this case as in that of the sailor deprived of such food,—the systemic affection also in both, points to a poverty in the blood as its proximate cause, but here the similitude ends, the wanting blood element is not alike in the two instances—the systemic fault is by no means *identical*, and however beneficial a resort to vegetables might be in the one case,

---

* It may be here remarked, that in many instances of so-called scorbutic disorder, much mischief is frequently occasioned by the indiscriminate and inordinate employment of fruits and vegetables:—the popular notion that such diet is necessarily beneficial, cannot be too strongly deprecated.

no one, we imagine, save a homœopath or a lunatic, would exhibit such treatment in the other.

Secondly; dermatological nomenclature does not, even to dermatologists, convey any certain and distinct signification,—a given name used by one gentleman to *distinguish* a certain form of eruption, may represent to the mind of some other gentleman an entirely unlike form of cutaneous disorder, so that, even for conversational purposes, these appellations are in many instances insufficient to convey any definite information.

Different authors vary in their description of the physical characters, progress, and situation of the several genera; not slightly, but, frequently, to such a degree, as to be at issue respecting the very nature of the so-called primary lesion presented by the genus in dispute,—*yet these primary lesions are the distinguishing characters which are supposed to constitute the ground work of the grand and fundamental division into orders.*

"The Porrigo Scutulata, or ring-worm of the scalp, has "given rise to great difference of opinion as to whether it "is a pustular or vesicular disease, and whether the "pustules or vesicles are at all essential to the disease. "Willan, Bateman, Biett, and the older writers, class it "among the former; some of the French writers,

"especially M. Cazenave, among the vesicular. Dr.
"Neligan considers herpes to be the true ring-worm;
"and Dr. Burgess regards this form as the result of
"normal irritation of the bulbs of the hair. When such
"eminent dermatologists differ, I cannot be expected to
"be able to decide. I can scarcely doubt, after the
"examination I have made, that there is a form of ring-
"worm, the element of which is a vesicle, but this does
"not prove that a pustular eruption may not assume
"this character. Dr. Burgess's description differs equally
"from that given by Bateman and that by Negligan." *

Again at page 624,—"Notwithstanding the opinions
"of Willan, Bateman, Alibert, Biett, and others, of the
"pustular character of porrigo favosa, it seems clearly
"established now that this variety at least is of a vegetable
"nature. It is true that Dr. Mahon considers it a morbid
"secretion of the subaceous glands, and Drs. Bennett and
"Burgess and Mr. Erichsen as a tubercular disease; but
"the researches of Schonlein, Gruby, Remak, Corrigan,
"Robin, etc., seem to have pretty well set the question at
"rest."

"Dr. Thomson, after full consideration of these diseases,
"decided not only on retaining the distinction drawn by

---

* Dr. Churchill's "Diseases of Children," p. 619.

"Willan and Bateman between porrigo favosa and
"lupinosa, but also concluded that the evidence in favour
"of the pustular origin of these diseases and of Pseutula
"was sufficient.

"This opinion is not, however, shared by many
"observers, who have classed together porrigo favosa and
"lupinosa as being different shades of the same disease,
"have denied their pustular origin, and have separated
"the affection, which, according to circumstances, is
"termed favus dispersus or confertus, from porrigo
"scutulata altogether. Still very considerable differences
"exist in the use of terms. Thus, 'porigo scutulata'
"is, according to Gustav Simon, a synonyme of favus
"confertus; 'porrigo lupinosa' is a synonyme of favus
"dispersus; while the herpes tondens of Cazenave is
"not referred to as at all allied to porrigo scutulata,
"although many facts imply that both appellations
"designate the same disease. Erasmus Wilson refers
"porrigo favosa to impetigo capitis, and terms porrigo
"lupinosa, favus dispersus. He gives the phrase porrigo
"'scutulata conferta' as a synonyme of favus confertus;
"while, in another part of the work, the porrigo scutulata
"of Willan appears as synonymous with his trichoses
"furfuracea, or Tinea tondens, viz., the herpes tondens of

" Cazenave, the herpes capitis of Neligan." " Examples
" of such varieties of nomenclature might be adduced in
" great numbers, if there were any object to be gained
" in so doing."*

" Among the causes of obscurity which attend the study
" of complaints of the skin, there is hardly a greater than
" the multitude of terms which are applied to them ; and
" if the various diseases had been represented by a chess
" board, and the names well shaken together in a bag, had
" then been emptied upon the board, so that several names
" should fall indiscriminately on each square, there could
" not be a greater confusion than reigns over the nomen-
" clature of these disorders."†‡

Thirdly ; the frequent impossibility of deciding as to
which form mentioned in the nosological tables, the case
under consideration is referable, forms no inconsiderable
impediment to the practical working of a method such as

---

* "Diseases affecting the skin," by the late Dr. Anthony Todd Thompson,
completed and Edited by Dr. Parkes, p. 430.

† Erasmus Wilson on " Healthy Skin."

‡ Mr. Wilson originally adopted a *Physiological* basis for his arrangement,
but, says Dr. Belcher (Neligan's Skin Diseases, page 11), " In consequence of a
more extended experience, Mr. Wilson abandoned this classification, and in the
fourth edition of his work, published in 1857, substituted for it what he terms an
Etiological classification," (the elaborate details of which would fill several pages),
whilst again, in the latest systematic treatise of that veteran (Student's Book,
&c., 1864-5), this etiological classification is in its turn set aside in favour of a
Clinical arrangement in twenty-two groups.

that which we have at present under consideration. Willan's system prescribes set rules of treatment for each several form of eruption bearing a generic appellation ; but it is of course utterly impossible to conform to such practice, in cases whose proper denomination does not clearly appear or cannot be discovered.

This untoward circumstance is by no means infrequent in its operation, nor does it depend altogether upon differences of opinion amongst dermatological authorities in regard to the physical attributes which, according to each, severally, should represent and constitute a peculiar genus or species. Where the constitutional peculiarities, habits, circumstances, and ages of patients differ ; where the duration and progress of the systemic fault may vary ; where accidental and fortuitous circumstances may interrupt, retard, or change the action of the morbific agency engaged ; where, moreover, the causes may be not only various in character, but complicated and concomitant, we can readily believe, that the resulting cutaneous symptoms will not, in any case, resemble too closely the figured or verbal description which may be put forward as typical of any genus.*

---

* The cases so freely published as examples of cures effected by this or that plan of treatment are, from the same causes, practically useless, save indeed, as proof, that the *name* of an eruption being given, the treatment—irrespective of difference in constitution or causation—varies only with the predilection of the author for one or other empirical remedy.

The physical distinctions between the primary lesions themselves, are but differences in degree and stage, all being alike, the result and evidence of abnormal cutaneous nutrition, (or as some would say of cutaneous inflam_mation), and, therefore, so running the one into the other, as to render it, in many cases, impossible to say to which form the eruption present, offers the greater or less resemblance.

Fourthly; the simultaneous presence of several more or less distinct forms of eruption in the same patient, resembles, somewhat closely, the last objection; it likewise shews, in a most conclusive manner, that the form assumed by the cutaneous symptoms is no safe guide as to the nature of the constitutional disorder;—the system may be assailed by several causes of disorder, or the systemic affection present may be complicated by the implication of several organs, etc., etc., but still the constitutional disorder is indivisible, nor can it give rise to different *kinds* of disease in any single tissue at one and the same time; although the *degree* of its morbific influence upon such, may be altered and various, in proportion to the previous healthy or unhealthy condition of the organ or tissue over which its action may be exerted. In other words, several forms of eruption being together present in the same individual, such are not, severally, the result of a separate and distinct constitutional cause, nor are there separate and distinct constitutional affections connected in any way with each—the systemic fault, complicated or uncomplicated in its nature as the case may be, is but one and indivisible, and the cutaneous lesions, however their characteristic appearances may differ, are all and equally related thereto. Even were it otherwise the disciples of Willan could not attempt to resort to a plurality of different remedies, supposed by their method to be requisite for each of the several forms of eruption which might be presented in the case; whilst it is equally plain that a single plan of treatment which will be beneficial and curative for the whole, must be such as will restore the constitutional integrity of the system, and, if so, afford conclusive evidence that the form of eruption is no guide as to the plan of treatment which will prove successful. *The form of the eruption is no criterion as to its constitutional cause: the indications for successful constitutional treatment are not dependent upon the form assumed by the cutaneous symptoms!*

" The characteristic symptoms of diseases of the skin " may be mixed up together, and we often find many " different elementary lesions co-existing, especially in " acute cases. They are often *attended* by general

" symptoms, particularly those of more or less severe " irritation of the air passages and intestinal canal."*

Fifthly; the alteration of the form of eruption from time to time in the same case gives evidence that a similar constitutional disorder or local irritation is capable of producing, according to the duration of its action, different forms of cutaneous disease; so that the practitioner who is early in attendance on the case may see one form of eruption, whilst a later examination would reveal another form. Yet the cause to be removed, whether constitutional disorder or local irritation, is of the same nature at both periods ! !

" Again, the diseases which commence with one generic " character, are liable occasionally to assume another, in " the course of their progress :—thus, some of the papular " eruptions become scaly, and still more frequently pustular, " if their duration be long protracted; the Lichen simplex " and circumscriptus, for instance, sometimes pass into " Psoriasis; the Lichen agrius and Prurigo formicans are " occasionally converted into Impetigo; and the Prurigo " mitis is changed to Scabies. Moreover, it frequently " happens, that the characteristic forms of eruptive disease

---

* Burgess, p. 7.

" are not pure and unmixed, but with the more predominant
" appearance there is combined a partial eruption of
" another character; thus, with the papular Strophulus,
" with the rashes of Measles and Scarlet Fever, and with
" the pustular Impetigo and Scabies, there is occasionally
" an intermixture of lymphatic vesicles. And lastly, the
" natural progress of many eruptions is to assume a
" considerable variety of aspect, so that it is only at some
" particular period of their course that their character is to
" be unequivocally decided. Thus in the commencement
" of Scabies papuliformis and lymphatica, the eruption is of
" a vesicular character, although its final tendency is to the
" pustular form; and, on the contrary, in all the varieties
" of Herpes, the general character of the eruption is purely
" vesicular: yet, as it advances in its progress, the inclosed
" lymph of the vesicles acquire a considerable degree of
" opacity, and might be deemed purulent by cursory ob-
" servers. In like manner, the original pustular character
" of some of the forms of Porrigo is frequently lost in the
" accumulating crusts, the confluent ulcerations, and the
" furfuraceous exfoliations, which ensue, and which conceal
" its true nature from those who have not seen, and are
" unacquainted with, the whole course of its advancement.

" These circumstances constitute a series of natural

" impediments to every attempt at a methodical arrange-
" ment of cutaneous diseases. But it is more philosophical,
" as well as practically useful, to compromise these
" difficulties, by retaining in the same station the different
" appearances of a disease, in its different stages and
" circumstances, when our knowledge of the causes and
" remedies, as well as of the natural progress and termina-
" tion of it, is sufficient to establish its identity, than to
" separate the varying symptoms of the same disorder, and
" to distribute the *disjecta membra*, not only under different
" genera, but into different classes of the system, after the
" manner of Prof. Plenck. Such was the method adopted
" by Dr. Willan; and, although it may sometimes diminish
" the facility of referring individual appearances to their
" place in the nosological system, yet it greatly simplifies
" the classification, as well as the practical indications to
" which it conducts us." *

VI. and VII. The next two objections which we have
to offer to this system are, we consider, insuperable, and
may conveniently be considered in conjunction. That one
and the same cause may give rise to various forms of
eruption, is in itself a good and tangible objection to a

---

* Bateman.

system, which professedly advocates a special plan of treatment for each and every different form of affection to which the skin is liable, and more especially so, when such diversities exist in the appearances of the several disorders as, according to this method, to separate them most completely, the one form from the other.

Again, the fact that a similar cutaneous eruption may depend upon very many, and dissimilar causes and constitutional conditions, would, we should imagine, convince the most obtuse intellect as to the necessity for a corresponding diversity in the remedial measures requisite for the cure of individual cases; would prove that no routine prescription could benefit all, or even the majority of patients, subject to such similar eruptions, and point with no little force to the general disorder, in contradistinction to the local symptom, as *the* affection to be considered and removed.

" The same exciting cause will produce different kinds " of cutaneous disorder, in different individuals; thus, " certain substances, which suddenly derange the organs " of digestion, sometimes produce Urticaria, sometimes " Erythema and Roseola, and sometimes even Lepra and " Psoriasis; yet each of these shall retain its *specific* " character, and follows its peculiar course; thus also

"certain external irritants will in one case, excite the
"pustules of Impetigo, and in another, the vesicles of
"Eczema."*

"To what are we to ascribe the great diversity of
"physical characters which the different eruptions, not
"exanthematous, assume? We recognise and describe
"some as *papular*, others as *pustular, vesicular, sqamous,*
"and *tubercular;* but we are forced to acknowledge that
"we are ignorant of the peculiar changes in the functions
"and general condition of the system whence these
"diversities originate. It is, nevertheless, true, that in
"every condition of the habit, *originating or associated*
"with diseased states of the skin, the capillary vessels are
"the organs morbidly affected; and, according to the
"degree of this change from their normal action, the
"diversities in the physical characters of the eruption may
"be traced. The importance of becoming familiar with
"these diversities need not be insisted upon; they are the
"chief sources whence a diagnosis can be formed, *and from*
"*which*, in many instances, we must derive our information
"of the general constitutional derangement with which
"they are associated."†

---

* Burgess.
† Thompson, by Parkes. Opus cit. p. 187.

"Tho propriety of separating, in description, those
"cutaneous eruptions which aro consequent on the action
"of 'the poison of Syphilis, has long been admitted.
"Abstractedly, their derivation from a specific cause,
"and practically, their treatment by special methods, are
"sufficient grounds to justify such a separation. Syphilitic
"eruptions may assume tho form of any of the cutaneous
"eruptions, which do not spring from specific causes.
"They may present the physical characters of roseola or
"erythema, of various papular, vesicular, pustular, scaly, or
"tubercular eruptions, such as lichen, eczema, impetigo,
"ecthyma, rupia, lepra, psoriasis, lupus, &c. ; but they
"never assume the forms of the true exanthemata, or of tho
"other eruptions which spring from specific agents. The
"eruptions of variola, scarlatina, typhus or typhoid
"fever, can never be imitated by the effect of the
"syphilitic poison. Possibly, an eruption owning another
"specific cause may happen to develope itself in tho
"system of a person who has become tainted with the
"syphilitic diathesis, and may bo moro or less impressed
"by the presence of tho pre-existing constitutional
"disease." *

---

* Ibid. p. 352.

"Experience shows that diseases of the skin may be
"caused by what we call, for want of a better name,
"critical influences; nature thus sets up a salutary
"deviation towards the skin. As to the cause of the
"*special* form which cutaneous disease may assume, we are
"completely ignorant; we cannot tell why the exciting
"cause should in one case produce a pustule, in another a
"vesicle, in a third a papule, yet it is to this obscure point
"all our efforts should be directed, for on it probably
"depends the secret of the precise seat of cutaneous
"diseases." *

VIII. The form presented by an eruption is no
criterion as to its cause, and, consequently, is practically
useless to direct us in our choice of remedies. Yet, the
practice in cutaneous affections rests entirely upon the
hypothesis, that the form, situation, and physical peculiarities presented by an eruption are always sufficiently
well marked to allow of its differential diagnosis,
and to decide as to its appropriate denomination;
such diagnosis when correct, being considered as an
adequate guide and indication to a suitable plan of treat-

ment.*† No honest practitioner, however, will be found to deny the fact, that remedies, so indicated, very generally fail, sometimes aggravate the disorder, or that after a long series of trials with different reputed remedies cases will be found, not seldom, to resist all such efforts, and at length be given up as " obstinate " and " incurable."

The following is from *The Lancet*, August 15th, 1863; similar queries are constantly to be met with in the pages of the several Medical periodicals. The inferences which may be drawn therefrom are obvious enough.

"TREATMENT OF ECZEMA."

"To THE EDITOR OF 'THE LANCET.'—Sir, may I ask "through your coloumns what can best be done for a

---

* " Describing the individual eruptions of the skin I have spoken of the treatment adapted for each." Neligan's Skin Diseases, by Belcher.

+ In a clinical lecture " On two contrasting cases of acute eczema," Dr. Handfield Jones says, " There can be no question that the chief curative agent on both " occasions (the case relapsed after about six months from the first period of "treatment) in the above instance was arsenic." The second case, previously without benefit, treated with arsenic, rapidly recovered under simple and non-specific treatment. Dr. Jones mentions that " the appearance of the diseased "skin in these two instances was closely similar, it might be said identical. No " dermatologist would have hesitated to pronounce them both examples of the " same disease and in the same stage. . . . But (says the Doctor) were they " indeed the same disease ? I presume that a chemist who had two perfectly " clear colourless solutions before him whose composition was unknown to him, " would not consider them identical if they reacted differently with the same test. " Some of the ingredients might be the same, but there must be some material " difference between the two fluids. So it is, I believe, with cases of eczema like " those I have detailed to you ; they are apparently similar but not really and " completely so ; and, unfortunately, not so in that which is of far more impor-

"troublesome case of eczema of sixteen years' standing
"which has hitherto, with the exception of temporary
"relief, resisted all treatment? Has any fresh light of
"late years been thrown on the treatment of this worrying
"complaint? Perhaps some of your numerous readers
" and correspondents may have successfully fought it, and
"would not object to give (briefly) the benefit of their
" experience, and much oblige,

A READER OF ' THE LANCET,' FOR THIRTY-FIVE YEARS."

Systematic writers are compelled to acknowledge that in
any given form of eruption whose possible causes are both
numerous and dissimilar, no one plan of treatment will
always or even frequently be admissible ; this does not, however, prove to them, as it should do, the utterly impracticable
nature and inutility of differential distinctions : constrained

---

" tance, viz., their therapeutic requirements. . . . . That in our second
" case the disease was materially increased by the administration of arsenic there
" can be no doubt, and I am anxious that you should always bear in mind the
" possibility of this occurrence, and not look on this remedy as one which is in
" the least a specific for skin diseases. It is, I believe, nothing of the sort, its
" *modus operandi* is tolerably intelligible, and, like all our good remedies, or
" almost all, it is capable of doing harm as well as good. When I give nitric acid
" or tr. ferri muriatis, or quinine in an æsthenic bronchitis, I have the same
" fear before my eyes that I have when I give arsenic for the cure of a cutaneous
" eruption. I fear lest they should chance to act as irritants and not as tonics,
" or as I have previously put it, that they disturb the nutrition of the tissue
" instead of toning the vessels and nerves. . . . In conclusion, take the
" following as a fair inference from the foregoing experience. Do not suppose
" that diseases are uniform entities ticketed (so to speak) in books with their
" appropriate remedies, and that all you have to do is to find out the name in
" order to have the treatment. This may do for counter practice and sundry
" and various pathios, but not for rational medicine."—*The Medical Press and
Circular*, June 19, 1867.

to confess the causes variable and uncertain, they yet maintain an ideal speciality, and so, as one plan of treatment will obviously not suffice, they offer some score from which their followers may take their choice, or wh ich severally are recommended by different *authorities*.

Is this not a good and sufficient explanation why cutaneous affections are intractible, and so frequently beyond the skill of the practitioner? Is it not evide ntly the natural result of a system which ignores and o verlooks the *cause* of the disorder in anxiety after a minut e differential diagnosis of its *effect?* Placing implicit reliance on the unnatural system in which they have been educated, the disciples of Willan first determine the form and gen us to which the case under consideration bears an outward resemblance, and this *momentous* question being decided to their satisfaction, proceed at once to adopt the plan (or rather one of the plans) of treatment, which their method lays down as orthodox and necessary, for the particular form of cutaneous lesion to which they have, with so much judgment, allotted a local habitation and a name.*†

---

* It is necessary, in order advantageously to study the diseases of the skin, to classify them, *and thereby obtain a comprehensive* view of them (?), and a notion of their natural affinities, as well as to be able to distinguish one from another, and *to give a name* to any case of disease that presents itself."—*Hillier on Skin Diseases* Lond. 1865.

† " The diagnosis can generally be made by means of the objective phenomena or physical signs alone, without any assistance from the history or the subjective

Can these gentlemen forget how often diseases, bearing a similar outward appearance, originate and are perpetuated by entirely opposite morbific agencies? Or can they fail to observe the various and dissimilar forms and appearances to which one and the same exciting cause is capable of giving rise?

What after all do these different orders, classes, and genera signify?

How do the papulæ differ from the vesiculæ, or these again from the pustulæ or squamæ? Are they not all examples of the abnormal nutrition (inflammation) of the cutaneous tissues?—alike in being effects of disordered capillary action, and differing only in degree and stage. Again, what practical benefit can be derived from the multifarious divisions into which authors have succeeded in separating and sub-dividing these different form?

What signifies it to know that each and every little variety of appearance, of form, or of location, justifies the use of a distinct appellation?

---

symptoms. The history and the patient's own sensations will prove useful *in the way of confirmation.*"—*Hillier,* p. 29.

"Having diagnosed the case accurately, the question of treatment may then be profitably entertained."—*Hillier,* p. 31.

"As an aid to diagnosis, his (Willan's) system of classification is convenient, *but as a guide to treatment it is of no use*"!!!—*Hillier,* p. 9.

Will the fact that *Purpura* is purple, that *Erythema* is red, that *Miliaria* is like a millet seed, or that *Scabies* has a tendency to provoke scratching, help to elucidate the nature or the causes of these affections?* or, in any way, tend to point out the proper method of treatment which may be requisite? Certainly *no*,—each system of classification and every plan of treatment depending upon the appearances or situation of the cutaneous symptom, is void of all scientific character whatever, and calculated to accomplish no end save the mystification of the ignorant, and the disappointment of both patient and practitioner.

We may be met with the objection in favor of differential diagnosis, that the exanthemata (Small-pox, Measles, Scarlatina, etc.,) are pretty constantly accompanied by eruptions, which when present are more or less characteristic and pathognomic in appearance and in progress;—truly, from the specific nature of the poison to which such diseases are severally due, it is only what

* " Every form of elementary lesion (papulæ, squamæ, vesiculæ, maculæ, &c.) is met with amongst the syphilides; but for purposes of treatment it is infinitely more important to know that a disease is syphilitic than to know whether it is papular or squamous; the diathetic condition must be treated: all the local phenomena, however different they may be from each other, will in this manner be got rid of."—Hillier, page 11. Why this exception in the case of syphilitic diathesis? Reason tells us the same rule is applicable and essential in *every* case .

might be expected that such should be the case; but it must be borne in mind, that unlike all other causes of constitutional disturbance, these, from peculiarity in the poisons engaged in the causation of the general disease, run each a definite course,—and that the resulting symptoms *therefore* are also definite *to a like extent.* But were Small-pox and the other constitutional affections of the same kind always accompanied by their own several and peculiar cutaneous symptoms—which by the way we know they are not—such fact would not, nevertheless, furnish one single argument in favor of the practical value of differential objective diagnosis ; it is not the characters of the several eruptions in the cases alluded to that furnish satisfactory indications for their management; it is a knowledge rather of the peculiar nature and action of their causes (the zymotic poisons), by which our treatment is governed—moreover, were it otherwise, such argument could in no manner apply to diseases referable to general and non-specific causes. In syphilis, we have an equally specific poison, yet the resulting cutaneous symptoms present the greatest diversity of character, and this simply because the natural action of the syphilitic virus is irregular and indefinite, presenting, therefore, in its cutaneous symptom, similarly various

appearances. The Syphilitic poison is not capable of
giving rise to eruptions similar in progress and appearance
to those depending upon *another specific poison whose
action and duration is constant and defined*, but, with this
exception, there is no form of cutaneous eruption which
may not arise in consequence of its action.

This difference between Syphilis and the before mentioned eruptive fevers is readily explained—the virus of
Small-pox, and the other similar specific and zymotic
poisons giving rise to general diseases which are usually
accompanied by cutaneous symptoms, have a rapid course
of action, and enjoy, so long as they obtain, the mastery of
the system:—it is not so in constitutional syphilis, the
syphilitic poison, besides being irregular, is slow in action,
and gradual in progress; usually extending over months
and years, and not rarely lasting during the whole period
of life, its influence is thus frequently exerted in connection with some other morbific agency, or its own individual
action is modified by its duration, by the treatment
to which it may previously have been subjected, by the
age, occupation, habits, or constitution of the patient;—
influences which have no time and less scope to operate in
the case of the exanthemata.

From diseases of the gravest import to instances of

slight and ephemeral disorder of the stomach, we have that little organ, the tongue, offering to our notice many and various alterations from the natural appearances which it presents when the body is in health, and such abnormal peculiarities, *together with the other symptoms and history of the case*, are truly to the physician both instructive and valuable. But imagine some inventive genius of the Willan school to map out these unnatural appearances according to their several physical and objective peculiarities, to divide them into orders, genera, and species, "after the manner of the Botanical arrangement of Linnæus," to allot to each a separate, euphonious, *perhaps meaningless* cognomen, and then gravely to enumerate as the symptoms peculiar to, and accompanying this or that alteration which may be presented by the tongue, such as are peculiar to the constitutional cause of the lingual disorder,—be it typhus, hydrophobia, dysentery, or a broken leg. If to this be added a system of special treatment for each separate form of lingual "disease" mentioned in such arrangement, irrespective of the nature of the accompanying symptoms in individual cases, and we have an excellent comparison to the systematic method of Dr. Willan and his admirers.

Finally; the following extracts leave us no room to

dou bt as to the special line of practice, and the supposed specific nature of the remedies, which are represented as requisite for the cure of eruptive disorders. Indeed, it would be impossible for eruptions to maintain their status as peculiar and distinct diseases without the prevalence of an opinion in favour of their specific medication. That they should particularly demand certain special remedies is compatible only with the supposition that they are due each and all, to special morbific agencies—we quote authorities sufficient to prove the contrary fact, viz., that they are dependent on the ordinary causes of diseases generally, and we know by experience, and it is in accord-ance with every reasonable inference, that they are likewise amenable to the curative influence of ordinary remedies exhibited with regard to their constitutional connections.

"The constitutional means employed in the treatment "of diseases of the skin are extremely various. They "comprise blood-letting, purgatives, alkalies, acids, anti-"monials, preparation of sulphur, sudorifics, and, finally" "the tincture of cantharides and preparations of arsenic "or mercury, which evidently act in a direct manner on "the skin." *

---

* Burgess.

"*Diseases of the Skin have been long submitted to a*
"*particular line of treatment;* viz., the use of bitters, and
"of remedies containing sulphur, which seems to have ex-
"cluded all others; within the last few years, however,
"several remedies of great value have been discovered,
"but careful observations were wanted to determine their
"real value, and *the cases to which they are applicable;*
"in supplying the latter knowledge, M. Biett has rendered
"a most important service. He was the only physician in
"Europe who has made *a complete series of experiments on*
"*the treatment of cutaneous diseases, with different reme-*
"*dies;* and it is a matter both of surprise and regret
'that many of the results which he has obtained should
"have been published by persons who conceal the source
"whence their knowledge was derived." *

"Local excitants are of various kinds and often very
"useful; they seem to modify the vitality of the skin.
"They comprise vapour baths and douches, alkaline baths,
"sulphureous baths of every kind, lotions or ointments
"containing mercury, sulphur, iodine, &c. *When speaking*
"*of the particular treatment of each disease,* we shall con-
"sider these preparations more fully." †

---

* Ibid, p. 16.
† Ibid, p. 17.

Notwithstanding the number and variety of the therapeutical aids above alluded to, it happens curiously enough, that certain authors, after arranging the various forms of eruptive disorder in accordance with Willan's method and their own peculiar fancy, and subsequent to a laboured description of, and distinction between the said genera and species, discover in some one or two favourite and so-called "specific" remedies, the cure for the majority, if not for all these varying affections. One advocates the Turkish bath, another mercury, and a third arsenic as *the* remedy for each and every chronic eruption of an "obstinate" and "intractible" character—and this, too, no matter what its cause, its connection, or the constitutional condition of the person affected.

Thus, a well-known author, after enumerating, classifying, and describing the multitude of genera and species mentioned by Willan, tells us, that in all chronic cases, the principal desideratum is a knowledge as to whether the case is, or is not, due to syphilitic contamination :—If syphilitic, this gentleman recommends a mercurial treatment—if not, in his opinion, there is no remedy equal to arsenic. That mercury is requisite in all, or even in any case, because of its syphilitic character, is a question open and undecided; however, to those (we are not amongst the

number) who consider it requisite and beneficial in
syphilitic affections generally there is nothing remarkable
in its employment as a remedy for differing forms of eruption having one common (syphilitic) origin ;—but what are
we to say to this wholesale employment of arsenic, in cases
acknowledged as dependent upon or connected with almost
every variety of constitutional derangement ? A system,
so self-evidently empirical, might reasonably be expected
to obtain a justly merited contempt—such, however, is not
the case,—the systematic method universally acknowledged, special plans of treatment for each differing form of
eruption have been found impracticable and useless—and
so, as any powerful therapeutical agent must, necessarily,
be beneficial and curative in *some* proportion of the cases
where it is experimentally exhibited, universal "specifics"
have come to find favour, in preference to the varied, but
none-the-less routine, plans of treatment recommended by
the more orthodox systematic writers.

"On the Medicinal use of Arsenicated Mineral Waters."
—*Lancet*, August 1st, 1863 :—

"While fully admitting that the whole subject of the
" physiological action and medicinal employment of this
" powerful substance requires careful and extended investi-
" gation, there cannot, I think, be any doubt that in many

"chronic intractible diseases arsenic is one of the most
"potent remedies at our command. *All* writers on
"diseases of the skin bear concurrent testimony to its
"value in the treatment of *lepra, eczema,* and other still
"more loathsome forms of cutaneous disease, and one in
"particular lauds its 'almost omnipotent influence' over
"the non-syphilitic forms of *the* malady, ascribing its
"occasional failure to the exhibition of the remedy in too
"large doses, and at intervals too distant. Mr. Erasmus
"Wilson, in treating of lepra, says that he places the
"greatest reliance on arsenical preparations; but in
"ordering the artificial solutions of this metal admits the
"necessity of explaining to the patient the symptoms
"which call for the suspension or omission of the medicine,
"and alludes to the precautions requisite to guard against
"irritation of the stomach by its ingestion. And in the
"most recent contribution on the subject, the lectures of
"Dr. M'Call Anderson, the same homage is paid to the
"superior efficacy of arsenic in the treatment of eczema.
"As regards this class of complaints I believe that the
"experience of most practitioners is to the same effect."

That arsenic, cantharides, and similar remedies, are most
potent agents is undeniable, but it should be remembered
that in exact proportion to their potency are they contra-

indicated for indiscriminate use. They all possess or are followed by well marked physiological changes, but can these effects, by any process of reasoning, be made to appear as suitable or curative in dissimilar and even directly contrary conditions of the system? The therapeutical agent, the physiological action of which is not calculated to prove beneficial in a given case, is, necessarily, and exactly in proportion to its potency, calculated to produce mischief, and it is deeply to be regretted, that the prevailing system adopted in these eruptive affections should lead to and sanction the use of special remedies, and, more particularly, that such pseudo-specifics are generally to be found amongst the most dangerous articles in the *materia medica*.

To *try* the effects of a powerful agent—be that agent a drug, wet sheets, or the Turkish Bath, simply from preconceived opinion that it *should be* beneficial in any disorder bearing a given name, cannot even be looked upon as a scientific experiment, and unless its physiological action be in accordance with the constitutional indications or systemic condition, is not, however cautiously prescribed, without corresponding danger.

Cutaneous disorders rarely call for heroic remedies, nor is it ever justifiable or safe to have recourse to such solely on account of the physical and objective symptoms presented by an eruption. We must direct our treatment to the eradication of the cause before we can hope to remove the effect, not forgetting that obstinate affections of the skin may be removed temporarily at the expense of subsequent and lasting injury to the constitution. We do not intend or wish to decry the medical *use* of arsenic in suitable cases, accompanied or unaccompanied by cutaneous symptoms, but we do most firmly and unhesitatingly lodge a protest against the *abuse* of this mineral, and particularly against its exhibition to fulfil a supposed indication furnished by the form and appearance of a cutaneous eruption. Practically conversant with its use, we have also witnessed the injurious and sometimes dangerous effects attending its abuse; we have seen many cases where its exhibition has miserably failed to afford benefit to the affections of the skin which were supposed to indicate its employment; and more than this, we have had the satisfaction to see recover, under the use of ordinary and simple remedies, administered with a primary regard to the *cause* of the local symptoms, *and the nature of the systemic fault*, many cases supposed incurable, or which had resisted arsenic and other kindred " specifics " prescribed by gentlemen well accustomed to their employment. Not only so—we have learned from experience that in *no case* is the exhibition of arsenic *necessary* :—that simple remedies will, if scientifically prescribed, in all cases susceptible of cure, restore, with at least equal certainty, the integrity of the skin; without danger to the constitution, and, in fact, with benefit to the general health : and it is in consequence of this knowledge, confirmed in some thousands of cases, that we have for some years past entirely discontinued its employment.

Because the ordinary medicinal dose of arsenic is about fifty times less than that requisite to kill, it is argued that " It is, therefore, not only a safe medicine, but an un-
" usually safe one. For, let us apply this test to other
" medicines. Take the average dose of calomel to be four
" grains : multiply this by fifty and you have two hundred
" grains for a dose ! one-fourth of which would be fatal
" to most persons in one dose. Take two grains of opium
" for a medicinal dose. This multiplied by fifty would
" amount to one hundred grains, one-fifth of which has
" often been fatal, being equivalent to an ounce of
" laudanum, of which two drops have poisoned an infant.
" Or let us suppose that fifty times the ordinary dose of
" any medicine *not poisonous* be swallowed. Few patients,
" if any, would survive the experiment. Fifty ounces of

"Epsom salts, thirty drachms of magnesia, three ounces of
"rhubarb, a quart of castor oil at a draught, or two quarts
"of *black* draught!   Nay, who would survive twenty-five
"drachms of sal volatile, or even fifty wine glasses of
"brandy for a dose?   Who would not prefer to risk his
"life on two grains of arsenic rather than on any one of
"these unheard-of doses of domestic medicine?   We might
"pursue this subject and shew that, whereas it requires
"fifty times the medicinal dose of arsenic to poison a
"patient, four or five times the medicinal dose of *any* of
"the active medicines in common use would prove poi-
"sonous in a few hours.   *Ergo,*—and who can avoid the
"conclusion?—arsenic in medicinal doses is safer, about
"eight or ten times safer, than almost any other medi-
"cine!"*

Alas! this magnificent specimen of mathematical reasoning is entirely inappropriate and useless—the proposition
omits all mention of the fact, that arsenic, unlike the
domestic remedies, IS CUMULATIVE ; so that the system of
the patient who is subjected to a course of this remedy
must, sooner or later, become saturated, and well may he
be thankful, if he be one of the "49 out of 50" in whom

---

* Hunt, " on the Skin. '

" a slight degree of conjunctivitis takes precedence of the more grave affections which indicate an over dose."

Amongst the accidents enumerated as liable to occur from the use of arsenic are "conjunctivitis," "swelling of the limbs or features," "irritation of mucous membrane," "purging," "gastritis," "desquamation of cuticle," "general inflammatory state," "flushings," "head-ache," "exhaustion," "restless nights," "sinking," "giddiness," "palpitations," "mental agitation and alarm," etc., etc.*

Now, if arsenic be really cumulative in its nature, if moreover, the system requires to be kept saturated with the medicine to the point or verge of danger (as in favorable cases indicated by sight ophthalmic inflammation), it matters little whether such saturation be effected by gradually increasing doses, suspended upon the manifestation of "unpleasant" symptoms, or whether slight conjunctivitis be induced, and kept up, by small continuous or diminished doses: the only practical difference between the two plans of exhibition, seems to be, that by the latter method, the point of saturation and danger is, by continu-

---

* " Arsenic is considered to be a tonic to the general system, but it certainly is not so to the digestive apparatus ; for, according to all testimony, extraordinary care has to be exercised in its administration in order that it may not injuriously affect the mucous membrane of the whole *prima via.*—Weedon Cooke on Cancer, p. 106.

ing the medicine in diminished quantities, always maintained; whilst to settle the question as to whether this
"almost omnipotent" remedy is truly "as harmless as
milk" we take the liberty to quote the following very
forcibly illustrative passage: " Beginning with five minims
"of Fowler's solution, three times a day, and continuing
"that dose steadily until the conjunctiva or tarsi became
" slightly affected; then reducing the dose again and again
"as the CUMULATIVE action became apparent in the state
" of the *tarsi*, I have VERY RARELY observed any of the
"more formidable results in the above catalogue of
"symptoms." And again, "If all subjects were equally
" susceptible of the action of the medicine it would be not
"only safe but advantageous to begin with at least twenty
" or thirty drops of Fowler's solution for a dose, this being
"the average dose borne without injury. But as we do
"not know what a patient will bear, different individuals
"varying considerably in their tolerance of arsenic, we
"begin with a moderate dose, say five minims three times a
"day. Now, a person taking this dose three times a day,
"will have taken in ten days, just 150 minims, and *the*
" *effect of this on his system will be the same as that of*
" *taking* 150 *minims (minus the quantity excreted in ten*

" *days*) *at a dose.* That this hypothesis is very near the
" truth *I have demonstrated in hundreds of cases."* ! ! ! *

But besides its cumulative, and, therefore, undeniably
dangerous nature, there is another, and equally valid
objection, to the almost universal and indiscriminate
employment of this pseudo-specific. *It very frequently
fails*, even after long and persistent employment, (no
matter in which way exhibited), to effect its supposed
curative property: and even in cases were eruptions *are*
benefited by its action, it too generally is found, that such
amelioration of symptoms is only temporary—the so-called
disease returning as soon as the "remedy" is omitted.
It may *for a time* combat an effect (the symptomatic
eruption), but unless accidentally, and rarely, it of
necessity fails to remove the constitutional cause upon
which such effect depends.

Of 140 cases of "psoriasis" or "lepra vulgaris" (these
scaly forms of eruptions it should be borne in mind are
par excellence supposed to be most fitted for, and susceptible to the greatest benefit from, the arsenical method)
treated with different preparations of arsenic at the St.
Louis Hospital of Paris, M. Emery thus writes : "38 only

---

* "On the Medicinal action of Arsenic in Cutaneous and other Diseases," by
Thomas Hunt, Esq., *Medical Times*, 1850.

have gone out apparently cured, after two, four, six, eight, and even fifteen months of treatment. Within six months, six of these had returned to my wards. Within eighteen months, I had readmitted twenty-two. I have never again seen the other sixteen."

So much for the value of this wonderful remedy ! ! !

Those who advocate the use of arsenic in almost all the forms of eruption mentioned by systematic authors, especially when such forms become chronic and resist their other remedies, must either view any given case as a purely local disorder of the skin, or as connected with some general disorder of the system at large. Now it is evident enough that a purely local affection of the skin, would with equal if not with increased readiness, yield to a more direct and local system of treatment, and one, which according to the dictates of reason, would be much more applicable than the resort to a constitutional remedy of so potent a nature:—for it is not easy to conceive, how a remedy acting through the system can remove an external and purely local disorder, without producing some change in the constitutional condition of the patient :—the pre-existing condition being one of health, any change or alteration therefrom must necessarily be of an injurious character. If, on the other hand, these cases are regarded as due to disorder of the general health, the disorder accompanying cutaneous affections being also notoriously of very various kinds,—how is it, we would ask, that in these same systemic or constitutional affections when severally unaccompanied by cutaneous symptoms, arsenic is not equally popular and in equal requisition?

Willan and his followers have given to symptoms the character of distinct diseases, whilst their want of success in treatment has led to a complicated system of empirical practice and specific medication, no less uncertain and equally unsuccessful. This is proved by the writings of both special and general practitioners, which not only abound in contradictions, but, ingenuously enough, allow the short-comings of their system, and even, frequently, acknowledge the true and symptomatic nature of eruptions.

The late Mr. Plumbe, in his work on the skin, than which no better exists in our language, thus very emphatically expresses an opinion adverse to objective diagnosis and to all systems of classification based upon such foundation: " A classification of the external and ever-changing " forms of the accumulated secretion of disease on the " surface—one day a pimple, the next a vesicle, on the " third a scab, or crust, the fourth a falling scale, the fifth " a red spot! This might have served the purpose at the

"time for want of a better, but to pronounce it a better
" classification than one founded, whether with solid foun-
" dation or not, by its originators, on etiology or the
" *causes*, external and internal, of the cutaneous disease,
" would bo manifestly absurd! There was a grand and
" fundamental error committed by Plenck, and imitated by
" Willan and Bateman, Alibert, Rayor, and others, in
" classing cutaneous disease in the manner mentioned."
" To know the cause is surely to know more towards its
" successful treatment, than the practice of the eye on the
" ever-varying forms of external morbid secretions can ever
" promise." * Surely, "adoration of precedent " would
seem to bo inseparably connected with this subject, for
Plumbe himself, in this same book, published coloured
plates intended to represent and distinguish the different
forms assumed by eruptions: makes use of Willan's
nomenclature, although "homely phraseology is, perhaps,
far better : " and particularly notifies his "attention to
descriptive details."

Inconsistencies are to be found in the writings of almost
all gentlemen who attempt cutaneous diseases ; for if at all
acute, they must necessarily see and acknowledge the

---

* A practical treatise on *the* Diseases of the Skin, by Samuel Plumbe, etc., etc.
4th edition.

connection between the local affection and a general
disorder of the patient's health; whilst at the same time
so firmly fixed is the idea of the distinct and special nature
of "skin diseases" generally, that they all, when on the
subject of treatment, with one accord practically ignore
the symptomatic character of the eruption, and even,
frequently, altogether disregard that very constitutional
relationship which they had previously pointed out. * So
indeed it ever will, and ever must be, whilst medical men
respect the artificial system now in such general repute.
The real nature of cutaneous affections is patent enough,
but so long as special remedies and plans of treatment find
favour for the treatment of these cases, so long will
eruptive disorders continue to cast an unmerited reproach
upon scientific medicine.

No plan of treatment, if directed only to the cutaneous
symptoms, can offer any probability of permanent success
so long as the cause, whether local or constitutional, upon
which the cutaneous disorder is owing, remains in opera-

---

* "In all cases special attention must be directed to the influence of diathesis,
the gouty, the scrofulous, the rheumatic, the sanguineous, the syphilitic, the
dartrous, &c., call for their appropriate remedies irrespective of the kind of local
eruption; or, to put it in another way, the general treatment varies in the same
disease, according to the general aspect of the patient; all deviations from the
standard of health must be rectified before or in conjunction with the employment of *special* medicines." Dr. Fox on Skin Diseases (1864), page 20.

tion; wo have, therefore, an all-sufficient indication in the removal of the cause (tho constitutional affection or tho local irritant); tho which object successfully accomplished the local and symptomatic affection of tho skin will either disappear spontaneously, or yield readily enough to such simple and ordinary means as arc calculated to restoro tone to tho capillary vessels of tho skin; which then alone, and in consequence of lengthened disorder, may occasionally retard tho cure.

The following is from tho well-known lectures of Dr. Thomas Watson, Physician extraordinary to tho Queen.

" Both the scaly disorders, lepra, and psoriasis, require "tho same kind of treatment. I believe that external "applications arc of but little use. I have tried a good "many, and have lost all confidence in them, with tho "exception of tho warm bath. Whatever tends to improvo "the general health, will hasten the departure of these " eruptions. I believe that they sometimes depend upon "tho presence, or the generation, of an excess of acid in "tho system; and that they arc often to be cured by " alkaline remedies, I am sure. I have seen many cases of " psoriasis rapidly improve, and get ultimately well, under "full doses of tho liquor potassæ; from half a drachm to a " drachm, three or four times daily, in a glass of milk, or of

"water, or of beer, or of ginger tea. Another internal
"remedy from which I have seen manifest improvement
"result, is arsenic; given with the cautions, and in the
"doses, which I have more than once spoken of. These
"are the two remedies of which I have the most
"experience; but neither of them is infallible; and you
"will have to try many things in succession, for patients
"are very desirous of getting rid of the disfiguring eruption
"even when it does not interfere with their health or
"comfort. Now the Harrowgate waters, a strong decoc-
"tion of dulcamara, pitch-pills (and if pitch-pills, I should
"suppose *a fortiori*, creosote), tincture of cantharides, and
"the iodide of potassium, are remedies of some renown for
"these scaly diseases."

We find this gentleman just previously, p. 774, expressly
and very justly condemning the method which he nevertheless adopts, teaches, and so perpetuates.

"Under the general head of *cutaneous* diseases are
"included maladies of very different kinds, and of very
"different degrees of importance. Some are attended with
"fever, and run a definite course, and are often dangerous
"to life. Others are chronic, irregular in their progress,
"troublesome perhaps, and obstinate, and disfiguring, yet
"implying no peril to the existence of the patient. Some

" again are contagious, while many are not so. But before
" I enter upon any further account of these diseases, I wish
" to make you acquainted with the names by which the
" various morbid appearances presented by the skin have
" been known, since the time of Dr. Willan.

"That author—whose works have been augmented by
" Dr. Bateman, so that perhaps I ought to say *those*
" authors—divides cutaneous diseases into eight orders
"distinguished from each other solely by the appearances
" upon the skin. I shall omit the last of these orders, the
" order of *maculæ*, such as freckles and congenital spots
" and discolorations, because in fact these are not diseases
" at all. *   *   *   *   *   *   *

" Now it is very convenient, for the purpose of distin-
" guishing different diseases, and of describing them, to
" know these outward marks when you see them, and to
" use these names. But they form a very unfit basis for
" the *classification* of diseases. Maladies may usefully be
" classed according to their causes; according to their
" intimate nature; according to the general plan of treat-
" ment they may require. *But the superficial markings of*
" *disease have no definite relation to any of these heads.*
"Besides, a complaint which is papular to-day, may be
" vesicular to-morrow, and pustular next Saturday. Yet

" the classification most commonly followed in this country,
" and in France, is that of Willan and Bateman. Here we
" find collected under one and the same division, maladies
" which nature has stamped with broad and obvious marks
" of distinction: the febrile with the non-febrile; con-
" tagious complaints with those which have not that
" property: ailments that are local and trivial, with
" diseases of grave import and deeply rooted in the system
" at large. And, on the other hand, distempers which
" nature has plainly brought together, and connected by
" striking analogies and resemblances, this methodical
" arrangement puts widely asunder. I point out, without
" professing to remedy, these imperfections."

The above very forcible expression of an adverse opinion
with regard to Willan's system does not, however, prevent
Dr. Watson from describing a majority of the principal
genera in accordance therewith, whilst the following
passage from his work pretty conclusively acknowledges
his practical belief in objective diagnosis:—" If you look at
" the list of *genera* and *species* appended to the various
" works which treat exclusively of cutaneous diseases, you
" will find that they are exceedingly numerous. But these
" disorders differ widely in their relative importance; and
" the principles upon which their remedial management

" proceeds are not so greatly diversified as these 'tables of
" contents' might lead you to suppose. I have spoken
" pretty fully of the most serious and interesting of these
" maladies—I mean the febrile exanthemata; but I have
" no time left for pursuing in detail the host of chronic
" affections to which the human skin is subject. Nor do I
" much regret this. *To become expert in the diagnosis of*
" *these blemishes, and in curing such as are curable by our*
" *art, you must see them with your own eyes.* Verbal
" descriptions of their changeful characters are of com-
" paratively little service or interest. They are among
" the things that require to be ' oculis subjecta fidelibus."
" Even pictured representations convey but an inadequate
" notion of the morbid appearances they are designed to
" portray. The lecturer on skin diseases should have
" actual patients before him *to whose bodies he could*
" *point!!* *

" The truth is, that the various genera of cutaneous
" disease, as characterized by their external appearances,
" do not differ in the same essential degree, in which the
" diseases of organs of various structure differ from each
" other."†

---

* Vol. II., p. 932.   † Bateman.

Again, Dr. Burgess, whom we have seen to be an uncompromising advocate for Willan's system, "differential diagnosis," and special remedies, yet notwithstanding, is compelled to acknowledge the true relative connection between cutaneous disorder and constitutional derangement.

"In some rare cases, where the cutaneous disease is "slight and limited in extent, local measures may suffice; "but, generally speaking, a constitutional treatment is "necessary, for cutaneous diseases are almost always "connected with some derangement of the general health, "against which local remedies are powerless." *

"It is scarcely necessary to remark that every deviation "from the healthy condition of the skin, if it cannot be "traced to the direct influence of some external agent, such, "for example, as great solar heat, which produces Eczema "solare; or to dry powders and alkaline solutions, which "produce two distinct local forms psoriasis, and similar "causes, must be regarded as depending on some morbid "change, either in the function of digestion, or assimilation, "or secretion, or in some derangement of the general "constitution.†

* Burgess, p. 18.   † Thomson by Parkes, p. 186.

"Indeed, in treating these affections, we cannot hope
"for success without having constantly before us their
"intimate connection with the organic functions." *

"It is, indeed, impossible to form a correct idea of almost
" any disease affecting the skin, by the mere inspection of
" the eruption, however important the characters of the
" eruption may be in aiding our diagnosis; we must exert
" our observing, discriminating, and reasoning powers in
" connecting the external indications of the disease with
" the general condition of the system; and the fact, that
" it is only by treating the latter that the former can be
" removed, ought never to be forgotten." †

Language like the above might well have been penned
by the opponents of classification, but when such spon-
taneous and unpremeditated criticism is found in the
works of gentlemen who themselves indulge in classification and its adjuncts, it is more particularly apposite and
to the point;—in fact, no more forcible evidence of the
inutility and obstructive tendency of their method could
possibly be adduced;—if cutaneous diseases are so
generally connected with derangements of the general
health, and demand constitutional treatment,—if the

---

* Ibid, p. 187.   † Ibid, p. 189.

constitutional disorder gives rise to the cutaneous symptoms and requires our first consideration,—where is the necessity for all this classification, this minute attention to " differential diagnosis," and this assumption of a specific character for cutaneous symptoms, and for cutaneous remedies ? Nothing can be more obvious, than to follow the rational plan of treating the constitutional cause,—to ignore a system based upon puerile, irregular, and meaningless distinctions,—and to strike at the root and cause of the totality of the symptoms, rather than to direct especial attention to the one, which if the most evident, is also frequently the least worthy of consideration.

Indeed for the treatment of these cases we require physicians conversant with all forms and phases of disease, —men whose observations have not been confined to any single set of symptoms, but who from experience of all stages and forms of disease generally, will assuredly be the most competent to treat any and every form of symptom to which such diseases are capable of giving rise. The varied and innumerable complications and connections, which, of necessity, exist, in and between all disorders, renders it an utter impossibility, that an *exclusive* attention to any particular disease should fail to lessen that full appreciation of general and relative derangement, a proper consideration of which is so important in all cases. The practitioner who has had the greatest opportunity of witnessing and treating diseases accompanied by cutaneous disorder, will naturally, and with the proviso that he has proper views as to the symptomatic character of such local affections, be the best qualified to treat these disorders; nevertheless, however common may be the contrary opinion, cutaneous eruptions are in no wise suited as the object for *exclusive* study, and no amount of skill or aptitude in diagnosis, no extent of experience in the use of psuedo-specifics, can ever replace or compensate that due appreciation of their real nature which is an essential to their successful treatment.

The rational indications for the treatment of eruptions may be briefly stated to consist—in the removal of local causes of irritation, should such be present—and in the employment of such general remedies as are calculated to restore the healthy integrity of the constitution.

Those cutaneous disorders which are originated as purely local affections from the direct action of some irritant upon the skin, are comparatively rare; they may, however, and, if their progress is long unstopped they inevitably will, produce a sympathetic and corresponding constitutional disorder, the which, once set up, will again in turn re-act upon the local affection, until in time the reciprocal action of both—the local disorder and the constitutional affection —being unchecked, the patient will become the victim not only of a chronic and " obstinate " affection of the skin, but also probably will suffer materially in his general health.

Some of the more local forms of eruption are rendered peculiar from their being accompanied by the presence of certain animal and vegetable parasites, which if they do not always give rise to the eruptions which they accompany, at any rate by their presence produce an amount of local irritation, which maintains and aggravates the morbid action existing in the part affected. Moreover, as these living bodies multiply and increase, so also will the surface over which their influence is felt increase likewise; the skin become more and more engaged and disordered, or to use a common expression, the eruption will be found to " spread." There can also be little doubt that these affections are capable of communicating a similar disorder by their direct contact, or by the transfer of the parasite engaged from one individual to another.

Much has been written controversially respecting this peculiarity (the accompanying parasites); one author maintains the parasite to be the *fons et origo mali*—the

first great cause of the disease, and that it is always from
contagion that such affections take their origin; another
considers that the parasite results from a pre-existing
disease in the part affected; a third, that though capable
of communication from one individual to another, it nevertheless requires a peculiar condition, susceptibility, and
aptitude in the part to which the morbific agent is applied,
before such can exert its natural and injurious effects! —
the immediate contact of the surface of the recipient, either
with the previously affected person or with some article of
dress, etc., infected by him, is thought by some to bo
necessary; whilst others imagine that the nidus of the
disease may be conveyed in the atmosphere, and so affect
those who may be susceptible to its influence.

The above are a few of the contradictory opinions which
find favour with the profession. They might be increased
indefinitely were there any utility in so doing,—disquisi-
tions of this kind are however useless to the practical
physician,—if the parasite be present, if by its presence
it keeps up and aggravates the cutaneous irritation and
disorder, and if through its agency the affection is capable
of communication from one part to another, or from the
one patient to others,—it suffices practically to know how
to distinguish its presence, how to destroy the offending

agent, and how to restore the injury already inflicted, or prevent the subsequent operation of its influence.

One thing is certain in connection with parasite affections, and this applies more or less to the sources of local irritation generally—that whilst all may suffer from their action, they are more common, and attack in preference, the feeble, ill-fed, and less cleanly portion of the community,—a fact of much significance and great practical importance. That state of the constitution called strumous seems particularly favourable to their formation and perpetuation, so much so, that by some they are regarded as necessarily connected therewith,—no doubt in a large majority of cases a constitutional disorder is present, and that, not rarely, such systemic fault acts as a powerful predisposing cause:—but it must not be forgotten that every local disease, however set up, is capable of giving rise to a more or less severe form of general disorder— hence it is always advisable, and sometimes indispensable, to direct attention to the constitution, even where the cause, and the disorder directly resulting from its action, may seem apparently to be only of a simple and local nature.

In reality many apparently simple cases demand, even more than others where the systemic and general disorder is plain and distinct, a full enquiry into their constitutional relationships and the previous history of the patient. True such enquiries demand tact, discernment, and no small professional knowledge of diseases in general:—the practitioner may experience more trouble and difficulty in such inquisitions than is demanded by the mere ocular appreciation of physical peculiarities on the surface; but it is in such discriminating faculty or talent that the true physician excels, and from the insight so obtained into the personal peculiarities of his patient's constitution, it is that he may hope to find his practice both satisfactory and certain.

The local causes of cutaneous irritation and disorder are readily enough discovered from the history of the case, nor are they, with the exception of the parasites already mentioned, at all remarkable or peculiar. The requisite measures for their removal are equally obvious, and with regard to the use of outward applications when the affection

is not altogether local, we have only to repeat that they can be of little service, unless combined with such proper constitutional treatment as the case may demand, and must always, when required at all, be exhibited in accordance with the stage or degree of disorder which may be present in the cutaneous capillaries.

The chief use of external applications is, that they may, in suitable cases, when judiciously applied, afford relief from irritation of a distressing character, and by so doing husband the strength of the patient until such period as our other general and more important remedial measures may remove the constitutional cause; but still it is in an appreciation and proper treatment of the latter that we must rely for a permanently good result. The *abuse* of external applications is not uncommon : nothing is easier, in many cases, than to dry up an eruption of long standing or to cause its disappearance, but the patient, who under such circumstances imagines himself cured, too speedily is undeceived—the eruption may or may not return, but its retrocession will in all probability be followed by the implication of some internal, perhaps vital, organ, or the system at large be dangerously affected. Such practice cannot be too strongly deprecated. Further than this we need not expatiate,—the details of practice in every case must necessarily vary, nor can any set rules ever replace a correct professional judgment as to the present and individual requirements of the patient.

So also, and more especially with regard to the constitutional treatment—there can be no plan which will suit all cases, however in the main similar—the causes alike,—the

constitutional peculiarities of patients must differ—the
stages and circumstances vary—and a hundred things
interfere with any general or universal method. Moreover,
the constitutional causes of cutaneous disorder, direct,
indirect, and predisposing, are innumerable, so that even to
attempt to enumerate and enter into detail of them all, or
of even the major part, would in fact be tantamount to
writing a general treatise upon the principles of medicine.
Such is not the object of the present work, still less to
endeavour to particularize the several remedial measures
which each differing constitutional condition might seem to
indicate. The practice of Medicine is not to be learnt from
books—*they* can but point out the general principles which
should govern a scientific choice of remedies—practical
acquaintance with disease, together with a sound judgment,
must do the rest. To the well informed and practical
physician the foregoing pages will have sufficiently indicated the *principles* upon which a satisfactory practice
may be conducted, nor would any more lengthened
argument or explanation prove of service to the unskilful
or prejudiced practitioner. To the latter its very sim-
plicity will, doubtless, afford indubitable evidence of its
" unscientific " character, but to the former, and let us
hope, more numerous class, this will prove no obstacle, and

to such, therefore, we respectfully and earnestly commend a trial of the rational method—well assured from abundant and successful experience, of its infinite superiority to the system now in use.

FINIS.

G. HILL, STEAM PRINTER, WESTMINSTER BRIDGE ROAD.

www.ingramcontent.com/pod-product-compliance
Lightning Source LLC
Chambersburg PA
CBHW020324090426
42735CB00009B/1392